Housing
93

珊瑚在意

Coral Care

Gunter Pauli

冈特·鲍利 著

凯瑟琳娜·巴赫 绘

唐继荣 译

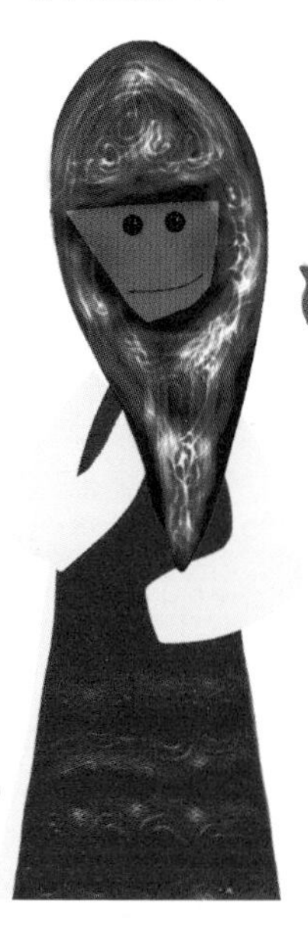

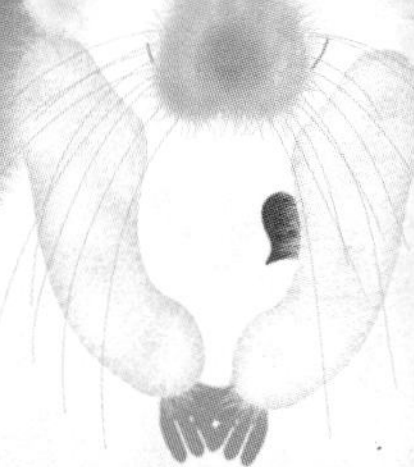

特别感谢以下热心人士对译稿润色工作的支持：

姜竹青　韩　笑　贾　芳　刘　晓　张黎立　刘之杰
高　青　周依奇　彭　江　于函玉　于　哲　单　威
姚爱静　刘　洋　高　艳　孙笑非　郑莉霞　周　蕊

目录

珊瑚在意 4

你知道吗? 22

想一想 26

自己动手! 27

学科知识 28

情感智慧 29

艺术 29

思维拓展 30

动手能力 30

故事灵感来自 31

Contents

Coral Care 4

Did you know? 22

Think about it 26

Do it yourself! 27

Academic Knowledge 28

Emotional Intelligence 29

The Arts 29

Systems: Making the Connections 30

Capacity to Implement 30

This fable is inspired by 31

褐藻一家沿着坦桑尼亚东北部的桑给巴尔岛海岸漂浮着，他们有很多抱怨。

“如果周边的人类继续用炸药捕鱼，他们将会毁灭珊瑚。”其中一个褐藻说。

A family of brown seaweed floating along the coast of Zanzibar has a lot to complain about.

"As long as people around here fish with dynamite, they will continue to destroy the coral," says one of them.

褐藻一家

A family of brown seaweed

海藻和鱼类在周围有珊瑚的环境下生长得最好

Seaweed and fish grow best with coral around

“用不着你告诉我！”一只珊瑚说，“我一直试图说服这些渔民，如果他们将我们都炸掉，将没有更多的鱼了。”

“难道他们没有意识到海藻和鱼类在周围有珊瑚的环境下生长得最好？”

“是啊。而且他们需要食物，不能只靠吃褐藻为生。”珊瑚说。

“You’re telling me!” says a coral. “I have been trying to convince these fishermen that if they blow up all of us, that there will be no more fish.”

“Don’t they realise that seaweed and fish grow best with coral around?”

“Exactly. And those people need food. They cannot live off brown seaweed alone,” says the coral.

“所以他们试图吃你们，珊瑚先生，那是非常无用的。”褐藻说。“任何人想咀嚼你，都会把牙崩碎。”他补充道。

“你看，我们都必须在一起生活。只要有我们珊瑚，周围就会有鱼类和海藻。”

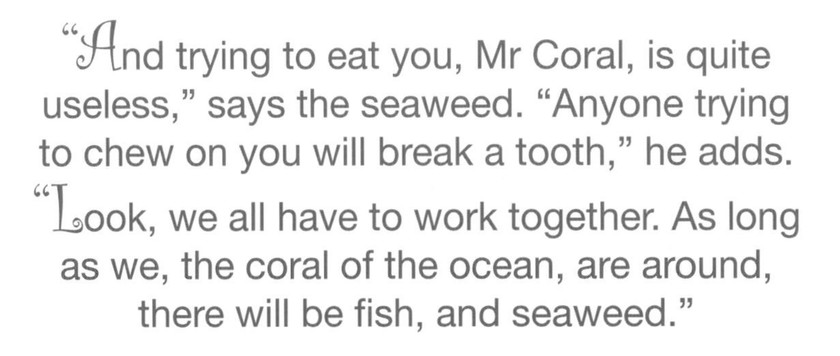

“And trying to eat you, Mr Coral, is quite useless,” says the seaweed. “Anyone trying to chew on you will break a tooth,” he adds.

“Look, we all have to work together. As long as we, the coral of the ocean, are around, there will be fish, and seaweed.”

把牙崩碎

Break a tooth

六周时间生长

Six weeks to grow

“只要有你们珊瑚在周围，我们就能免遭海浪的冲击，不会被拍打得很厉害。那样，我们会生长得更快、更大，因为海里已经有我们需要的所有食物啦。”

“你们要多久才能生长成熟呢？”

“我们从100克长到1000克需要6周时间。”

“And with you coral all around, we are protected from the rough seas and being buffeted around too much. We can grow big a lot faster then, with all the food already available in the sea.”

“How long before you are fully grown?”

“It takes us six weeks to grow from weighing a hundred grams to a kilogram.”

“这就是说，你们在6周时间里增重10倍！真是够快的！没有其他动植物能与此相比。”

“你看到过有些妇女在养殖更多我们这样的海藻吗？她们把海藻串在水中的绳子上，而在水中我们能获得大量的食物。当然，只要有充足的食物，同时有你们保护我们免遭海浪的冲击，我想我们甚至可能比竹子长得还快。”褐藻说。

“That means you pick up ten times your weight in only six weeks! That is really fast. There is no animal or plant that can match that.”

“Have you seen women planting more of us, tying seaweed twigs onto strings in the water where we can get loads of food? I think we grow even faster than bamboo, provided of course that there is an abundance of food for us — and that you protect us from the rough seas,” says the seaweed.

保护我们免遭海浪的冲击

Protect us from the rough seas

为孩子寻找食物

Find food to feed their children

“被别人需要的感觉真好，谢谢你！但是，我们该怎样对付那些使用炸药的渔民呢？我们应该请警察来抓住他们，并把他们送进牢房吗？”

“噢，对于饥肠辘辘而且只知道通过毁灭珊瑚来捕鱼的人来说，最好的警察、最快的船和最坚固的牢房都不能阻止他们试图为孩子寻找食物。”

“It feels good to be needed, thank you. But what shall we do about fishermen using dynamite? Shall we ask the police to catch them and put them in jail?”

“Well, if people are hungry and the only way they know to fish destroys coral, then the best cops, the fastest boats, and the strongest jails won’t keep them from trying to find food to feed their children.”

“你说得对！穷人正遭受痛苦，没有人希望挨饿。难怪他们没有耐心。”

“你肯定是一只非常有耐心的动物。”褐藻评价道，“你需要很长很长时间才能成熟，但爆炸在瞬间就能彻底毁灭你。”

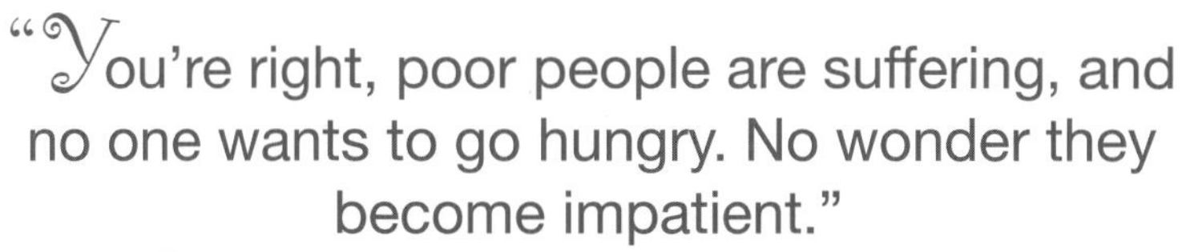

“You’re right, poor people are suffering, and no one wants to go hungry. No wonder they become impatient.”

“You must be a very patient animal,” remarks the seaweed. “It takes you ages and ages to grow to maturity, and in less than a second an explosion can completely destroy you.”

你肯定是一只非常有耐心的动物

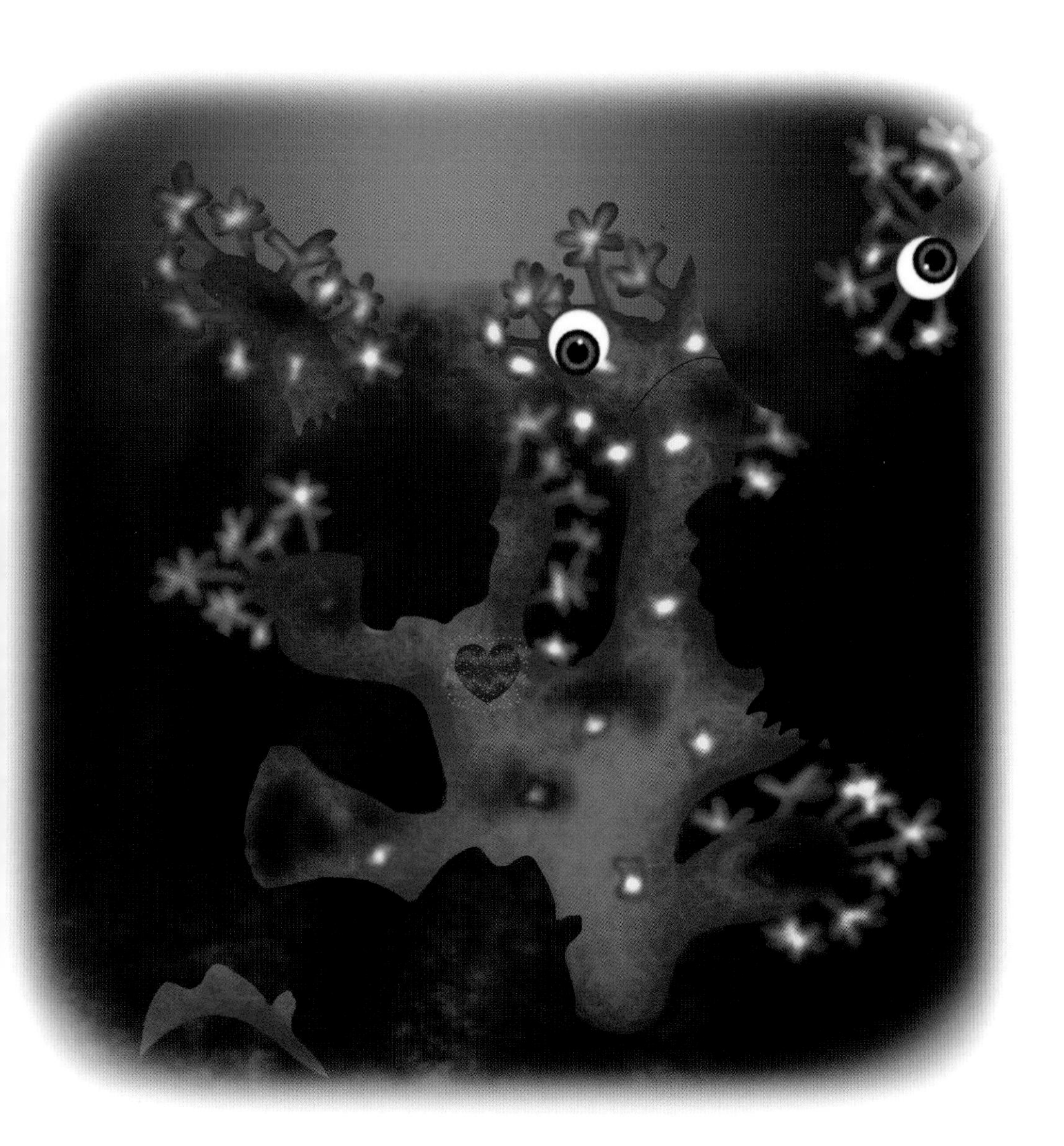

You must be a very patient animal

我们可以在你身上种蘑菇吗？

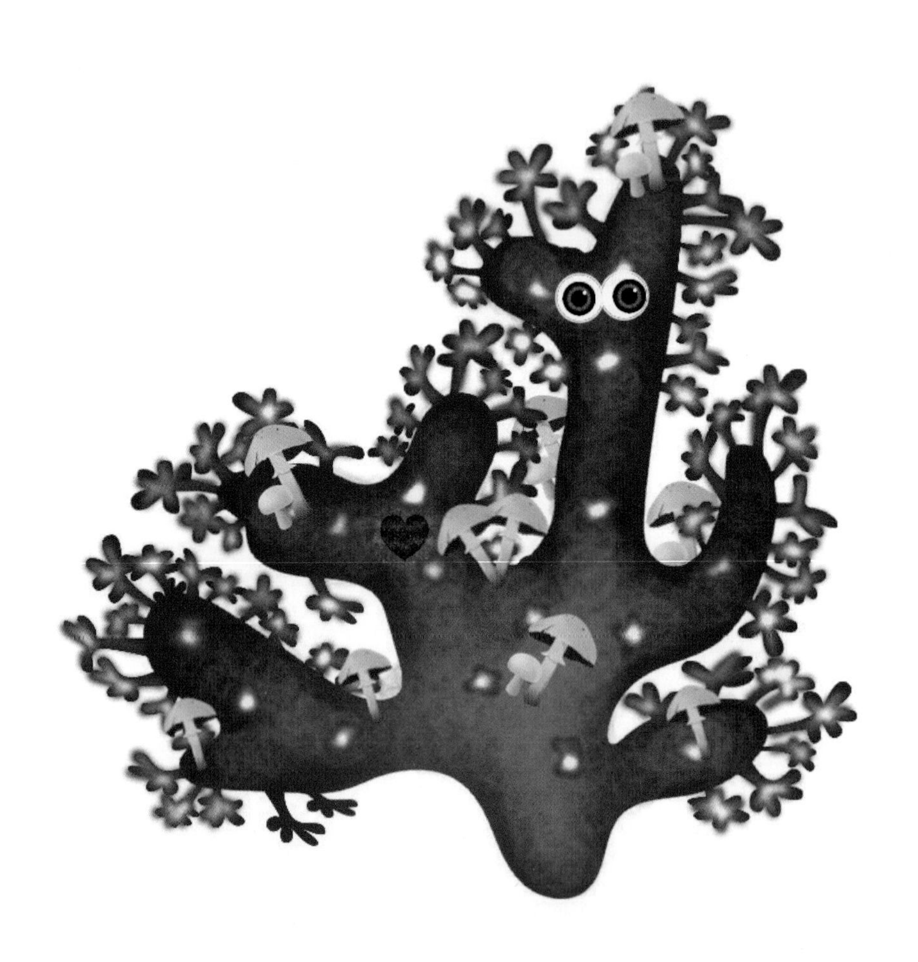

Should we grow some mushrooms on you?

“噢，我只是一个小小的珊瑚生物，也叫水螅体，需要很多我这样的个体才能创造出一块大型珊瑚礁来。虽然我可能活不长，很快死亡，但我们这个大群体存活的时间却非常长。只要我们周围的环境不被炸药、污染物或化学物品改变，我们的群体就能存活下去。”

“那就让我们设法确保在沿海生活的穷人有足够的健康食品。我们可以在你身上种蘑菇吗？”

“Well, I’m just one tiny coral creature, or polyp. It takes many of us to make up a large coral. While I may not live long and die quickly, our large colony lives for a very long time. That is as long as our environment is not changed by dynamite, pollution or chemicals.”

“So let’s find ways to ensure that poor people living along the coast have enough healthy food. Should we grow some mushrooms on you?”

“想尽一切办法去做吧！当人们把你夹入到饭碗中时，他们也能添加一些蘑菇！”

“这将让他们变得更聪明。”

“不仅更聪明，还更健康。你猜会怎样？在将来，甚至可以用你们海藻生产T恤！”

……这仅仅是开始！……

“By all means, do. Then when people add you to their bowls of rice, they can add some mushrooms too!”

“That will make them much smarter.”

“Not only smarter, but also healthier. And guess what? In the future one will even be able to use you seaweed to make T-shirts!”

… AND IT HAS ONLY JUST BEGUN!…

……这仅仅是开始！……

...AND IT HAS ONLY JUST BEGUN!...

Did You Know?
你知道吗？

In Asian cuisine, seaweed is added to soups, broths, stir-fries, and even wraps. It is used to wrap around fast foods like sushi. Seaweed is good for healthy gut flora and it also thins blood and removes heavy metals from the body.

在亚洲人的食谱中，海藻被加入到素菜汤、肉汤、炒菜甚至包装产品中，用于包裹像寿司这样的速食。海藻含有益健康的肠道菌群，也可以稀释血液，从身体中排出重金属。

Seaweed is the lungs of the ocean and an important harvester of carbon dioxide. Coral protects coastal zones from storms and erosion while providing a habitat for spawning and nursing fish.

海藻是海洋的肺，是重要的二氧化碳吸收者。珊瑚保护海岸带免遭暴风雨侵蚀，并为产卵和育卵的鱼类提供栖息地。

褐藻在日本被称为“昆布”。海藻中最重要的元素是碘，这是所有脊椎动物生存所必需的。

In Japan, brown seaweed is known as kombu. The most important element in seaweed is iodine, which is indispensable for all vertebrate life.

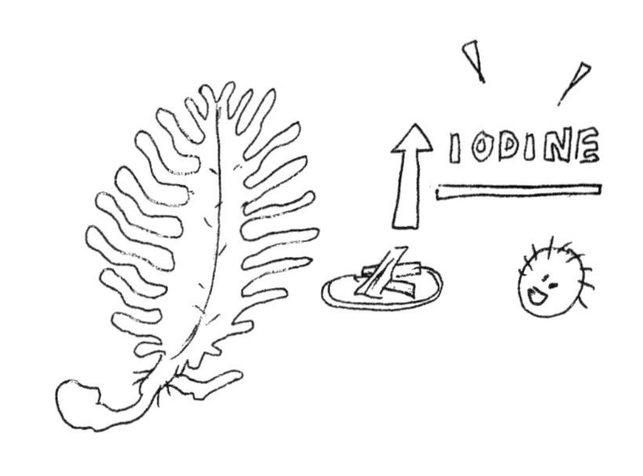

约有 5 亿人依赖珊瑚礁来获得食物、保护、建筑材料、收入，或开展旅游。在澳大利亚、菲律宾、伯利兹和美国，有十几个独特而重要的珊瑚礁地点被联合国教科文组织认定为“世界遗产”。

Approximately 500 million people depend on coral reefs for food, protection, building materials, income, and tourism. There are a dozen unique, important UNESCO World Heritage coral sites in Australia, the Philippines, Belize and the USA.

Coral reefs are under threat from global climate change, unsustainable fishing, and land-based pollution. About 20% of coral reefs have already been lost and 15% is under serious threat.

珊瑚礁正处在全球气候变化、不可持续的捕鱼和陆地污染的威胁之下。约 20% 的珊瑚礁已经消失，另有 15% 的珊瑚礁正受到严重的威胁。

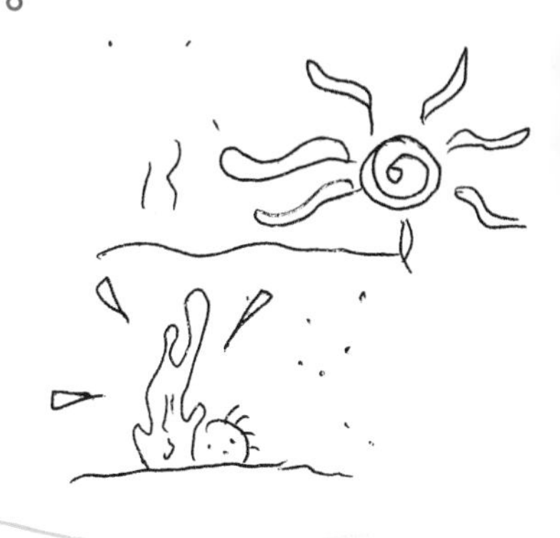

Coral reefs support 4 000 types of fish and 800 hard coral species. That is more species per unit area than in any other marine environment. Scientists estimate that there may be as many as a million undiscovered organisms living in and around coral reefs.

珊瑚礁支持着 4000 种鱼类和 800 种珊瑚的生存，珊瑚礁单位面积上拥有的物种数量比任何其他海洋环境都多。科学家估计可能有多达上百万种未被发现的生物生活在珊瑚礁里或其周围。

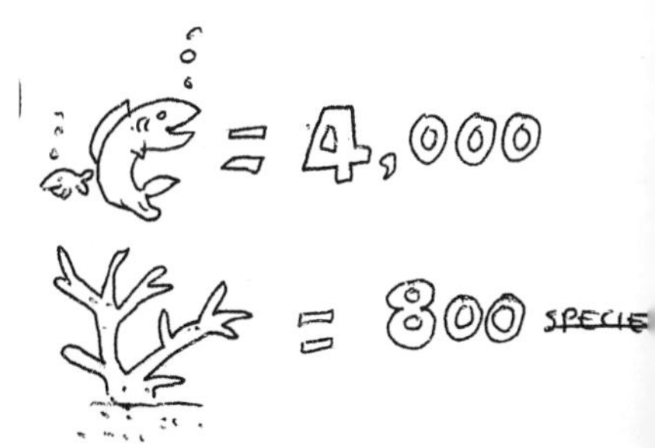

Most corals feed at night, on microscopic zooplankton, small fish, and organic particles in mucous film and strands. They use stinging cells to capture food.

大多数珊瑚在夜间进餐，取食微小浮游动物、小鱼和有机颗粒。它们用刺细胞来捕获食物。

Corals are colonial organisms, composed of hundreds to hundreds of thousands of individual animals called polyps that have stomachs that open at one end. The opening, called the mouth, is also used to clear away debris.

珊瑚属群居生物，由从数百个到数十万个的个体组成，被称为水螅体，其胃部在一端开口，也被用于清除体内残渣。

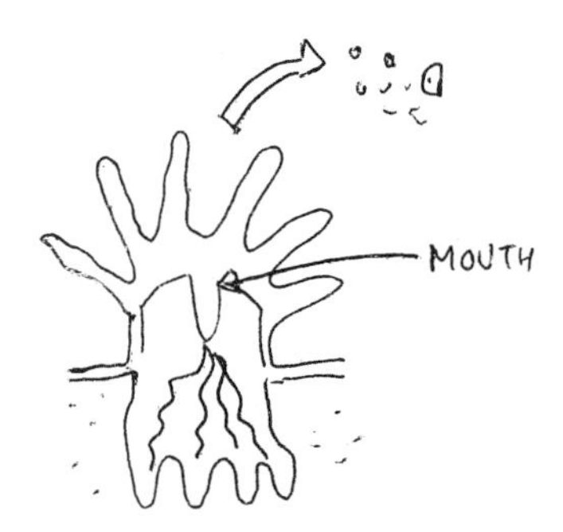

Would you use explosives to catch fish unless you were desperate?

除非处于绝境，否则你会用炸药来捕鱼吗？

Have you ever had a breakfast that included seaweed?

你吃过含有海藻的早餐吗？

How efficient is the production of food when it comes to you instead of you having to go find it?

可以直接获得而无需寻找的食物，它的生产效率如何？

When people are hungry and need to feed their children, do you think they would be prepared to break the law to get some food?

当人们很饥饿并需要为他们的孩子提供食物时，你认为他们会用违法手段获取食物吗？

Do It Yourself!

自己动手！

Have you ever had the opportunity to see a live coral? Here is what you can do to protect it: The less water we use, the less polluted water ends up in the ocean, damaging and killing coral. Even if you live far from the coral reefs, be aware that fertilisers – especially non-soluble ones – flow into the ocean and damage corals. Any kind of trash pollutes water and floats down to the sea, sometimes in tiny particles. When you buy fish, or go snorkelling, or stay at a beach resort, ask people there what they are doing to protect the reefs. When we plant trees, we help protect coral, as better tree cover reduces run-off and, consequently, the amount of tiny trash particles ending up in the sea. Do your part by spreading the word about protecting coral and learning how to propagate and plant new coral.

你亲眼目睹过活的珊瑚吗？下面告诉你怎么做才能保护它。我们使用的水越少，最终流入海洋并伤害或杀死珊瑚的污水就越少。即便你住的地方远离珊瑚，也要注意各种肥料（特别是非水溶性肥料）会流入海洋伤害珊瑚。任何一种垃圾都会污染水体并流入海洋中，有时以微小颗粒的形式出现。当你买鱼、潜水或在海滨度假时，询问人们会采取什么措施来保护珊瑚。我们植树也是在帮助保护珊瑚，因为更高的植被覆盖会减少地表径流，相应减少最终流入海洋中的微小垃圾废物颗粒。你也可以尽力传播保护珊瑚的知识，并学习如何种植新的珊瑚。

学科知识

Academic Knowledge

生物学	珊瑚是一个很庞大的动物家族，某些珊瑚需要8年才能成熟，生长速度最快的珊瑚每年可长15cm；红藻、绿藻与褐藻的区别；海藻是海洋之肺，产生可穿过消化道的非水溶性纤维，影响肠道菌群；珊瑚为在其中生存的物种创造生存空间，并提供全套的生态系统服务功能；巨型海藻每天生长0.5m，最终长达30~80m；海胆啃食巨型海藻，能消灭所有的巨型海藻，创造出海胆荒漠。
化学	炸药由硝化甘油、硅藻土和碳酸钠制成；海带是一种褐藻，富含维生素B_1、镁、铁、钙、维生素B_2、维生素B_5、木脂素、碘和褐藻素；巨型海藻用于获取碳酸钠，制造炸药；虽然用海藻种蘑菇成本太高，但海藻提取物（琼脂）可用来制作微生物的培养基。
物理	硝化甘油受到物理冲击就爆炸，然而当它被硅藻土吸附时，就能耐冲击；水下炸药的导火索由氧化剂供氧，不会在水中熄灭。
工程学	炸药主要用于开矿和道路建设；海藻生物精炼需要水，而水在实验室用来制作胶凝剂和培养基时只利用一次就被浪费了，但如果也用作灌溉就会非常高效。
经济学	海藻加工后的废水是很好的稻田肥料，为植物提供丰富的碘，这是它进入食物循环的途径；欧洲国家为向盐中添加碘的公司提供补贴，就是为了发展当地产业，保证碘能在食物链中循环。
伦理学	不要以暴制暴；采用毁灭性的方法获取食物，会在不久的将来降低你为家庭提供食物的能力；穷人是否应该因没有水和食物而失去耐心去寻找解决方案。
历史	捕鱼活动在4万年前出现，开始是徒手捉，后来发展出鱼叉、网捞、鱼钩钓、陷阱诱捕等方式；一直到1940年，南非都是最大的炸药生产国。
地理	在历史上，印度洋上的桑给巴尔岛是一个贸易港；菲律宾和印度尼西亚是世界上最大的海藻生产国；印度尼西亚拥有最丰富的海藻生物多样性。
数学	海藻在6~8周时间里增重10倍。
生活方式	整日暴露在太阳下时，需要保护皮肤免受损伤。
社会学	炸药使用具有争议性；当犯罪的根源在于饥饿和生存需要时，压制性方式的有效性如何?
心理学	一个人感到受欢迎，或是被社会需要并能作出贡献的重要性。
系统论	珊瑚礁提供各种各样的生态系统服务，如食物来源、生存空间、孵化育幼场所、保护海岸带免遭暴风雨侵蚀、药物来源、生物多样性，以及文化根基；珊瑚礁生态系统是珊瑚、藻类和许多其他物种的共生体。

情感智慧

Emotional Intelligence

珊　瑚

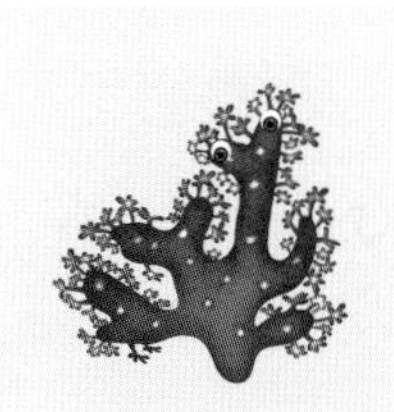

珊瑚正在经历一场危机，他非常清醒地知道人们迫切需要食物，而海藻不能独自提供足够的营养。他请求包括鱼类、海藻在内的所有生物共同努力。珊瑚有敏锐的观察力，对不容易理解的方面提出问询。他惊讶于海藻高水平的生产力，并认为养殖海藻可能是一个解决方案。珊瑚探究动用警察的压制性方式，但迅速懂得饥饿的紧迫性会迫使人们去做他们能做的一切以求生存。珊瑚对生命有一种沉思的态度，充分了解炸药、排放到海洋的化学物质和在水中阻碍阳光穿透的土壤颗粒所带来的挑战。最后，珊瑚提供了一些实用建议来帮助人们变得更聪明、更健康。

褐　藻

褐藻担心珊瑚的毁灭，不理解人们竟然可以毁掉自己未来的食物基地。他认识到周围有健康珊瑚的重要性，知道自己快速增重的能力，但也意识到以下事实：如果不是因为与珊瑚的共生关系，他不能如此快地生长，也不会这样高产。褐藻用简单而又清晰的逻辑进行推理，探究为什么饥饿的人在直到没有鱼或珊瑚时才停止使用炸药。褐藻接受人们需要除自己以外的更多食物的事实，并且表示单纯供应足量的食物是不够的，食物还必须健康安全。

艺术

The Arts

是时候看照片了。你将在互联网上发现大量美丽而形态各异的珊瑚照片。选择你最喜欢的3张，与也选择了3张照片的朋友分享。从这6张照片中选出你们共同发现的最契合的共生关系，并表现出珊瑚礁与海洋中、海岸带上许多物种共同生活的3张照片。现在请第三位朋友选择自己最喜欢的3张照片，并重复上述过程。

思维拓展

Systems: Making the Connections

珊瑚礁处于濒危状态，它们的衰退是由许多因素造成的：化学物品（特别是洗涤剂、漂白剂和肥料）导致珊瑚死亡；土壤侵蚀增加了水的湍流从而阻碍太阳光进入水中；虾场的创建、珊瑚被用于道路建设和旅游度假地的开发，也都促进了珊瑚礁的死亡。采用炸药不仅在一瞬间造成需要多年才能恢复的环境损害，而且也摧毁了人们赖以获得每日食物供应的基础。如果只是采取现在可用的技术来供应食物，那么不管当前生态系统能供应什么，都不足以满足每个人的基本需求。这就是为什么我们需要学会怎样与当地现有的生产方式结合，以及怎样增强生态系统的功能。我们需要利用丰富的生物多样性来增加产出，同时恢复生态系统的反馈环、乘数效应和共生关系。这里的挑战是时间。珊瑚礁的恢复需要很多年，但现在就很迫切需要食物。这促使我们要找出哪些物种生长得更快，哪种方法能更好地应对危机：这不仅要提供食物，而且要有就业机会和生态保育。海藻的种植是一个有吸引力的出发点，因为海藻是天然产品，容易种植和处理，而且在全世界都有广泛需求。目前，海藻主要被用作食物和肥料，但其应用范围明显超出这些产业。海藻种类繁多，生长迅速，有可能找到一种海藻养殖方式，使之对人类及生态系统发挥多重效益。

动手能力

Capacity to Implement

海藻营养价值高，容易种植。列一份能从褐藻、绿藻和红藻提取的产品清单。挑战自己，找到至少20种应用方式。然后，研究在你的国家谁在种植和加工海藻，谁在进口海藻及其目的。在现有海藻商业的基础上，找出还能用海藻做些什么，这样，你可以建立起大量的海藻用途组合。就像咖啡产业那样，我们还只是看到了海藻产业的刚刚兴起。

故事灵感来自

凯托·姆什杰尼
Keto Mshigeni

凯托·姆什杰尼在坦桑尼亚的高地长大，这里离海洋很远。尽管如此，他仍然发展出对海藻的浓厚兴趣。他从达累斯萨拉姆的东非大学毕业，获得植物学和地理学的学位，并辅修了教育学。他在美国的夏威夷大学获得植物科学的博士学位，并在最大的海藻生产国菲律宾开展博士后工作。他在海藻养殖上的创新方法引导印度洋上的桑给巴尔岛、奔巴岛和马菲亚岛等岛屿的沿海村庄成功地引进海藻产业，改善了农村妇女的生活。他在纳米比亚大学继续他的研究，出任学术事务方面的副校长，而且在亨蒂斯湾创建了海水养殖综合系统。目前，他是位于达累斯萨拉姆的胡贝特·凯卢吉纪念大学的校长。

更多资讯

http://www.hkmu.ac.tz/media/more/prof._keto_mshigeni_wins_aau_award_of_excelllence_in_higher_education_and_r

http://www.divenewswire.com/03/

http://www.ryandrum.com/seaweeds.htm

图书在版编目（CIP）数据

珊瑚在意 ：汉英对照 ／（比）冈特·鲍利著 ；
（哥伦）凯瑟琳娜·巴赫绘 ；唐继荣译 . -- 上海 ：学林
出版社，2016.6
（冈特生态童书 . 第三辑）
ISBN 978-7-5486-1056-4

Ⅰ . ①珊… Ⅱ . ①冈… ②凯… ③唐… Ⅲ . ①生态环
境－环境保护－儿童读物－汉、英 Ⅳ . ① X171.1-49

中国版本图书馆 CIP 数据核字 (2016) 第 125803 号

冈特生态童书
珊瑚在意

作　　者—— 冈特·鲍利
译　　者—— 唐继荣
策　　划—— 匡志强
责任编辑—— 程　洋
装帧设计—— 魏　来
出　　版—— 上海世纪出版股份有限公司学林出版社
　　地 址：上海钦州南路 81 号　　电 话／传真：021-64515005
　　网址：www.xuelinpress.com
发　　行—— 上海世纪出版股份有限公司发行中心
　　（上海福建中路 193 号 网址：www.ewen.co）
印　　刷—— 上海丽佳制版印刷有限公司
开　　本—— 710×1020　1/16
印　　张—— 2
字　　数—— 5 万
版　　次—— 2016 年 6 月第 1 版
　　2016 年 6 月第 1 次印刷
书　　号—— ISBN 978-7-5486-1056-4/G · 391
定　　价—— 10.00 元